❧ *To* ❧

❧ *From* ❧

101 Quick Tips
To Make Your Home
❧ S O U N D ❧
SenseSational

Books by Terry Willits

101 Quick Tips to Make Your Home
Feel SenseSational

101 Quick Tips to Make Your Home
Look SenseSational

101 Quick Tips to Make Your Home
Smell SenseSational

101 Quick Tips to Make Your Home
Sound SenseSational

101 Quick Tips to Make Your Home
Taste SenseSational

Creating a SenseSational Home

If you are interested in having Terry Willits speak to your
church, organization, or special event, please contact:

InterAct Speaker's Bureau
8012 Brooks Chapel Road, Suite 243
Brentwood, Tennessee 37027
Telephone (800) 370-9932
Fax (615) 370-9939

101 QUICK TIPS
TO MAKE YOUR HOME
SOUND
SenseSational

TERRY WILLITS

ZondervanPublishingHouse
Grand Rapids, Michigan

A Division of HarperCollins*Publishers*

101 Quick Tips to Make Your Home Sound SenseSational
Copyright © 1996 by Terry Willits

Requests for information should be addressed to:

ZondervanPublishingHouse
Grand Rapids, Michigan 49530

Library of Congress Cataloging-in-Publication Data

Willits, Terry, 1959–
       101 quick tips to make your home sound SenseSational / Terry Willits.
           p.   cm.
       ISBN: 0-310-20227-2
       1. Noise control.   2. Dwellings—Soundproofing. 3. Sounds.
    4. Christian life.   I. Title.
    TD892.W54   1996
    640—dc20   96-13343
                                                                                    CIP

This edition printed on acid-free paper and meets the American National Standards
Institute Z39.48 standard.

Edited by Rachel Boers
Interior Illustrations by Edsel Arnold
Interior design by Sherri Hoffman

Printed in the United States of America

96 97 98 99 00 01 02 /❖ QF/ 10 9 8 7 6 5 4 3 2 1

Better a dry crust with peace and quiet than

a house full of feasting, with strife.

Proverbs 17:1

Introduction

— ⚜ —

God has given us ears as antennas to tune into the world around us, and to provide the brain with a great deal of information about our surroundings. Sound has a powerful effect on our minds, emotions, and memories. It can warn us of danger, locate things, bring pleasure or pain, and soothe or stimulate.

From the songs we play to the words we say, the choices we make about sound can intensely affect our home's atmosphere. There is perhaps no greater reflection of the condition of our hearts and our homes than the sounds which fill our walls.

May the following tips encourage you to listen carefully to the sounds in your home. Cherish those that are pleasing, eliminate those that are disturbing, and add some that are soothing. God bless you as you bring harmony to your home!

terry.

101 Quick Tips
To Make Your Home
❧ Sound ❧
SenseSational

1

Listen up.

Use the ears God has given you to monitor the condition of your home and heart. Open your ears and mind to the sounds filling your home. Make every effort to enhance your home's atmosphere with pleasant sounds and to reduce those that are unpleasant and within your control to change.

2

Tune in to the outdoors.

$\mathcal{T}$he sounds of the outside world can be the most soothing sounds to our ears. Open your windows and listen to crickets chirping, rain falling, wind whispering, thunder clapping —

even the beautiful silence of a

snowfall.

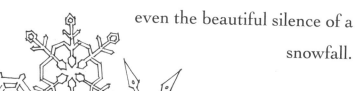

3

Come one, come all!

*M*ount a big iron bell on a wood post outside your home as a welcoming sign of liberty and love. Ring it to round up your children from playing in the neighborhood. Or have a simple bell you can take outside and ring as a "come home" call. Once they hear it, they'll come running!

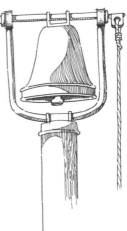

4

Cherish chirping.

Enjoy the sweet chirping of birds outside your home by placing a bird feeder or birdhouse near a kitchen window or any window that you view often. As the birds discover your friendly favor, your home will become a favorite place for them, and their singing will be music to your ears.

5

Hang charming chimes.

*H*ang a pleasant-sounding wind chime in a breezy spot on your porch, patio, or deck. It will make music with the whispering of the wind. Find a wind chime beautiful to your eyes and subtle to your ears.

6

Be friendly across fences.

Be a witness in the world in which God has planted you by showing love and care. Walk outside your front door and visit with your neighbors. Talk to them about what's going on in their lives. Courtesy is contagious. A friendly neighborhood starts with being a friendly neighbor.

Trickle in tranquility.

Enjoy the gentle sound of trickling water from a small, flowing fountain as it turns your garden, patio, or sunroom into a soothing oasis.

Many fountains contain pumps that recirculate the water, so no plumbing is needed.

8

Dingdong.

*D*on't overlook the ability your doorbell has for making a friendly first impression. The doorbell is usually the first sound guests will hear when visiting your home. There are many different doorbell styles, sizes, and sounds available. Choose a ring that suits your home and satisfies your ears.

Knock, knock.

Display a brass knocker on your front door that reflects your home's style. For an extra personal touch, have it engraved with your family name to welcome guests and let them know they are at the right home.

10

Tie one on.

*T*ie a lovely bell or string of bells to your entrance doorknob with a pretty piece of ribbon. The welcoming jingle will acknowledge when anyone is coming and going, and become a familiar greeting to loved ones stepping inside.

Know people's names.

$\mathcal{C}$all those visiting your home by name. The simple act of remembering others' names is a personal, caring touch that lets them know they are loved and that their lives are important. Ask your regular mail carrier, UPS deliverer, and service workers their names. Write them down and remember to address them appropriately when you see them. They'll never forget your home!

12

Shoo flies!

$\mathcal{T}$he buzzing sound of an insect zooming through a room can drive even the calmest souls crazy. Keep a flyswatter on a hook in a handy place. Tie a pretty bow on its handle and swat away whenever necessary.

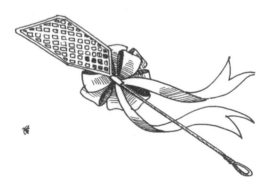

13

Arrive to instant atmosphere.

*S*et your stereo on a timer to come on before you enter your home at night. Or plug your stereo system into a wall outlet wired to a light switch. Walk in and flick the switch for instant atmosphere!

14

Honk your horn.

Make it a family ritual that whoever drives into the garage announce his or her arrival by honking the car horn. This little sound will give your heart a lift every time you hear it, knowing a loved one has made it safely back home again. As soon as you hear the honk, head for the back

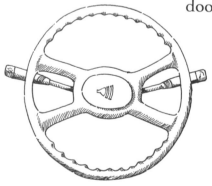

door to meet the person with a warm and loving greeting.

15

Set the mood
with music.

When family or friends enter your home, let the music set the mood for your time together. Select music appropriate for the occasion: Play peaceful, mellow music to wind down a long, hard day; upbeat music for a fun, lively party; calm, classical music for a touch of sophistication; jazz for an easy, carefree feeling.

16

Give a wind-down welcome.

*W*hen loved ones walk in the door after a busy day, give them time to wind down before stirring up a lot of conversation. Most people need to slow down and shift gears when they get home. Respect a loved one's need for a few private moments and he or she will be more likely to speak up when the time is right.

Quiet quirky noises.

While it's true that some quirky noises give a home its unique personality, many only give their homeowners a headache. Eliminate sounds that bother you in your home. Oil squeaky doors or drawers.
Fix noisy fans.
Replace creaky floorboards.

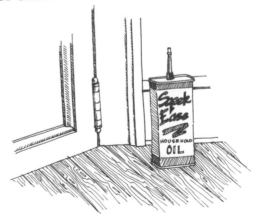

18

Express your love.

We can never hear enough of the three words "I love you." Don't miss an opportunity to let your family members know you love them. Tell them in the morning as they leave your home, in the evening before they go to sleep, on the telephone, in a note in their lunch box, or any time the emotion strikes your heart. These words of love will ring in their ears.

19

Sound off for safety.

*F*or maximum fire prevention safety in your home, install a smoke detector in every room that has a door that can be closed. Change the batteries annually to ensure safety. Many smoke detectors make a tiny beeping sound when their batteries need replacing.

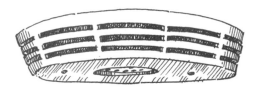

20

Wipe out whining.

Make a conscious effort to not whine or complain, and encourage your family to do the same. Constant complaining and whining signal an ungrateful heart and can drive others away or take them down with you. If you are struggling in this area, ask God to help you. Begin by thinking about and writing down your life's blessings.

Praise their presence!

*L*et others know you are genuinely glad to have them in your home by telling them so. Whether it be a child home from school or a visitor sharing a meal, they will appreciate knowing their presence matters in your home.

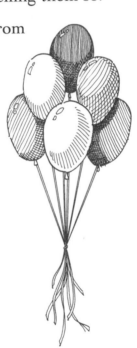

22

Hang up in a hurry.

*M*ake it a habit to get off the phone as soon as you hear a loved one come home. Kindly tell whomever you are speaking with that you don't like to be on the phone when your family walks in, and that you can continue the conversation at another time. This simple gesture will let your family, as well as your friend, know your priorities, and both will be blessed because of it.

23

Tick, tock.

*P*lace a beautiful clock with pleasant-sounding chimes in a prominent spot in your home. Most chiming clocks strike every quarter and announce the hour with their chimes. Whether it be a classic clock resting on the fireplace mantle or a stately grandfather clock filling your entrance, the familiar sound will ring "home" every time it strikes.

24

Silence heavy machinery.

*T*ry to avoid running the dishwasher, washing machine, dryer, or any noisy machinery after your family returns home at the end of the day. Appliances can be notorious noisemakers and can add to one's stress level. This simple gesture of silencing life's conveniences allows your family to hear only pleasant, familiar sounds when coming home.

Shop, look, and listen.

When shopping for appliances, look for models with sound control. Many dishwasher models offer quieter operation, and some food waste disposals are wrapped with insulation for less noise vibration. Ask to hear an appliance running before purchasing it.

26

Minimize vibrations.

To prevent your heavy appliances like the washer, dryer, or refrigerator from transferring additional noise vibrations to the supporting floor, place rubber pads under each leg or corner of the appliance. Allow at least two inches between the wall and your appliance, and between your washer and dryer, to eliminate banging noises.

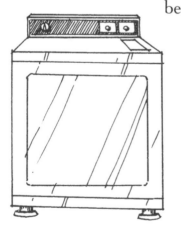

Nurture, don't nag.

Nurturing nourishes, or stimulates growth, in others; nagging annoys, or drives others farther and farther away, if not physically, at least emotionally. If you struggle with nagging, strive to make a request once, then drop it. Turn your frustrations over to the Lord, and try to focus on encouraging the ones you love.

28

Soak up the sound.

For a quieter home, choose furniture, fabrics, and finishes that absorb sound. Upholstered furniture and lined fabric draperies enhance quietness. Carpet is the best sound absorber for floors, but wood floors absorb

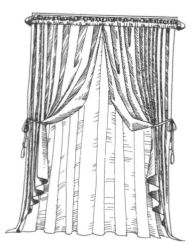

more noise than do vinyl, tile, or stone. Add area rugs to hard surface flooring to soften noise. Quiet walls by paneling or upholstering them.

Make a memory.

*U*se an audio or videotape player to record special moments in your home. Tape your child's comments after her first day of school. Interview a grandparent or parent about their past and your family history. Tape record your husband reading the Christmas story in front of a crackling fire. The familiar sounds of a loved one's voice will keep precious memories alive.

30

Get an earful
of reality.

*I*f you want a real earful of the sounds that fill your home, run a tape recorder unannounced during mealtime or any busy time in your home. Play it back to your family for a good laugh — and maybe to point out areas that could use improvement.

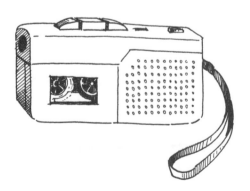

Add an aquarium.

*F*or a lovely, tranquil sound, set up an aquarium in a prominent spot in your home. The colorful fish swimming in their exotic surroundings will captivate your eyes, and the soothing hum of the motor running will calm your soul.

32

Avoid the television trap.

Television can be one of the greatest detriments to healthy relationships in a home. Turn off the tube and tune into each other's lives to bring calmness and consideration to your home.

Set up a super sound system.

$\mathcal{E}$xcellent sound can greatly enhance the atmosphere of your home, so buy the best sound system you can afford. Wire speakers into several adjoining rooms so you can experience great sound as you enjoy your whole home.

34

Blast off.

*F*or a great sound while viewing a video or television, hook up your television to your stereo speakers. The resulting sound will be fuller, richer, and louder. You'll feel like you're sitting in a theater or stadium!

Know about newsworthy news.

$\mathcal{K}$eep current on world happenings through one news medium; listen to or read about news just once a day. It's good to be informed, but there's no need to inundate your mind with unimportant, irrelevant, or destructive information.

36

Be a sport.

Celebrate the seasons by occasionally listening to a sports game being broadcasted. Whether on radio or television, swing into spring by listening to a baseball game or kick off fall by filling your home with the sounds of football. Even if you are not an avid sports fan, show love to those with whom you live by sitting down with them and getting into the sights and sounds of the game.

Round up the ringers.

$\mathcal{B}$egin a collection of pretty handbells with pleasing rings. Mix old treasures found at flea markets or antique shops with new ones discovered in gift shops or on family vacations. Display your beautiful bells, filled with memories, in a little spot looking for a charming touch of sound.

38

Be considerate and compromise.

*F*or the most pleasant atmosphere, it is best to have only one sound maker on at a time. Inevitably, however, where two or more are gathered, there will be times when different people have different sound preferences. In such situations, be considerate and compromise. Use a headset. Take turns selecting music. Close doors. Turn the volume down.

Save those special sayings.

When your child is young, keep a special book to record his or her inquisitive questions, hilarious sayings, and profound words. Chances are, you'll remember these sayings for a while as you relay them to others, but as life goes on, they'll be forgotten. Present the book to your child when he or she leaves the nest.

40

Snap, crackle, pop!

*I*f you have a fireplace, take advantage of the cozy sound of a crackling fire by lighting one often. Burn only dry wood that's been cut and seasoned for at least nine months. For an extra pop, occasionally toss in a pine branch or cone.

Make time for "couch time."

Make it a habit to set aside some time most evenings to sit down on the sofa and talk face-to-face with loved ones. Turn off distractions, put away projects and paperwork, and enjoy quality conversation.

42

Jingle your pillows.

Stitch a bowed bell on each corner of a square pillow. Every time it's tossed, it will jingle. Make holiday jingle pillows and give them as Christmas gifts.

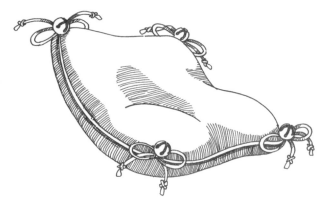

43

Retreat for rest.

$\mathcal{N}$o matter what our age, there are moments when we are tired, irritable, and unpleasant, and whatever comes out of our mouths is likely to be the same. Have a family understanding that when someone is

unpleasant, they be excused to the quietness of their bedroom for the rest they need. This will save many hurtful and unnecessary comments.

44

Love by listening.

Show others your love and concern by listening carefully to them and not interrupting their thoughts. Put down what you are doing and give them your undivided attention. You may hear something very important, or at least convey to them that they are important to you.

Say pretty please.

Set a good example for your family by using the word "please" as a part of your everyday vocabulary. "Please" softens a statement by turning it into a respectful request. People are much more likely to respond to a considerate comment than a command.

46

Have a humble heart.

We all blow it now and then, especially with those who we love the most. Keep a clean slate in your home by apologizing sincerely as soon after the incident as possible. Humble words like, "I'm sorry, will you forgive me?" can heal a broken heart and prevent walls of bitterness and resentment from building. Be an example to your children or spouse and humble yourself.

Lighten up and laugh.

$\mathcal{L}$ike a ray of sunshine, laughter lightens life's load by brightening outlooks and broadening perspectives. Lighten up and laugh in your home, especially at yourself. Rent a funny movie, relax, and enjoy your home life with those you love.

48

Skip the sarcasm.

Healthy humor helps, but unhealthy humor hurts. Sarcasm is cut-down humor, or humor at the expense of others; it may start out being funny, but it ends up being destructive. Do a sound check on the sarcasm in your home, starting with your own tongue.

Tune in
to your tone.

*K*eep a close check on the tone of voice in which you communicate in your home. Tone and volume say much more than words and can enrich or ruin a home's atmosphere.

50

Relive memories with music.

Keep a variety of different types of music that remind you of magical moments in your life. Buy a CD or cassette of the sound track to a fabulous movie or favorite musician of a concert you attended. During a romantic evening, play a special song that reminds you of your courtship. When reminiscing, play music from the era in which you grew up.

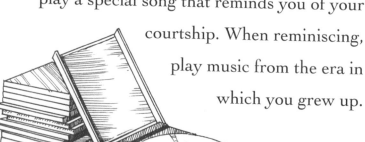

Make Christmas merry with music.

Celebrate the holidays by filling your home with your favorite Christmas songs all season long. Play one as you pull out your decorations. Turn one on when you curl up in front of a late-night fire. Praise God as you listen to songs that cause you to reflect on the miracle of Jesus' birth.

52

Keep it soft and subtle.

When entertaining, soft music is a wonderful backdrop to relax guests and fill in awkward gaps of silence. Though music can enhance the moment, make sure conversation can be shared without strain.

53

Say hello!

$\mathcal{Y}$our phone is your line to the outside world, so answer it with warmth and enthusiasm. Ministering to others means being kind and considerate, no matter how you feel or who is on the other line. If you can't handle it, don't answer it.

54

Communicate with convenience.

A cordless phone or a phone with a long cord will give you flexibility to do easy, quiet tasks, like ironing, while talking. Save time by making all your calls at one time. Simplify your life by never making a trip somewhere when a quick call will do.

55

Please leave
a message . . .

*H*ave an answering machine to take a
message when you're away from home or when
you simply do not want to be disturbed by the
outside world. There are many varieties on the
market; choose one that has the features you
desire. Make your greeting pleasant and
friendly. Keep your answering
machine in a convenient
spot where it won't
disrupt conversation
with family or friends.

56

Ring, ring.

If you're in the market for a new telephone, don't skimp on quality. Buy one with clear reception as well as a pleasant ring. Ask to hear a telephone's ring before you purchase it. A sudden, shrieking phone can unravel the nerves, so when you get it home, set the ring

volume as low as possible, but still at a level you can hear.

Turn on the timer.

When cooking, use a timer rather than continuing to open your oven door or pot lid. Set the timer for a minute or two before suggested cooking time; when it goes off, peek once for doneness.

60

Keep your cabinets quiet.

*B*ecause wood and laminate cabinets reflect kitchen sounds, putting dishes away can sound like a bull in a china shop. Soften the sound of clanging dishware by placing rubber or cork tile on the shelves of your kitchen cabinets. To

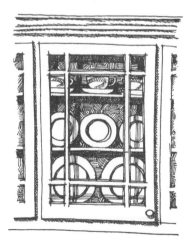

make closing the cabinets quieter, use soft, rubber or cork bumpers on the inside edges of doors and drawers.

Whistle away.

*B*uy a pretty teapot that whistles. When you need hot water, fill it up and listen — it will sing to you when your water is boiling.

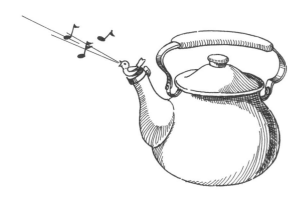

62

Listen while you work.

To make the most of time spent doing chores around the house, listen with your ears while you work with your hands. Upbeat music will motivate you to move quickly while cleaning. Christian teaching will make time spent ironing or cooking seem to go faster.

Have book ears.

*I*f you can't seem to find time to read, but want to enjoy a best-selling book, many are available on audiotape. Purchase an audio book at a bookstore, rent one, or check one out of a local library. Enjoy listening as you do tedious tasks around your home.

64

Cherish kitchen chats.

We can nourish others in our kitchen with encouraging conversation just as well as with delicious food. Kitchens are a natural gathering place for homework, paperwork, or just sitting and talking. Make the most of opportunities for casual kitchen chats. While you stir a hot pot of soup, someone may be warming up to open their heart.

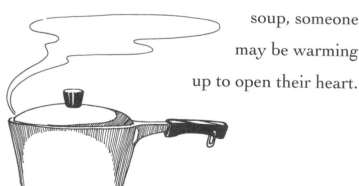

Ding for dinner!

*A*void the strain and sound of shouting throughout your home for others to come to dinner. Instead, have a familiar, pleasant-sounding dinner bell that signals dinner is almost ready. By the time loved ones wrap up what they're doing, dinner will be ready, and no one's nerves will be shot.

66

Let it ring.

$\mathcal{T}$ry to not answer the telephone during any meal, but especially during dinner. If the eating experience is to be calming and enjoyable, interruptions that can interfere with conversation and digestion should be avoided.

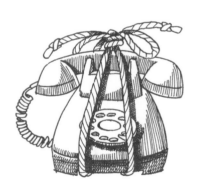

67

Make the most of mealtimes.

*T*ake every opportunity to sit down with others and eat together. If necessary, give young children a nap and a snack to tide them over so your family can eat at the same time. Try to keep the atmosphere positive: turn off the television, play peaceful music, and save heavy subjects for another time.

Make an effort to include everyone in mealtime conversations.

68

Bring on the bells and whistles!

*K*eep fun party noisemakers on hand for an instant sound celebration. Pull them out for birthdays, special announcements, and to ring in the New Year!

Serve foods that sound scrumptious.

*T*ry crispy chips, crunchy vegetables, popping corn, or sizzling steaks. Pleasant-sounding foods enhance the satisfaction and memory of a snack or meal.

70

Have an attitude of gratitude.

*B*egin every meal with a brief prayer of thanks to God for his provision of food. Take turns saying the mealtime prayer. No matter how few or simple the words, a prayerful pause will help everyone to reflect on God's goodness.

Thank you! Thank you!

*A*n attitude of gratefulness should carry over into our family relationships. Thank family members for doing the dishes, making their beds, and emptying the trash. Knowing their little efforts are noticed and appreciated will encourage them to keep helping.

72

House a feathered friend.

*E*njoy the sweet chirp of a canary or the smart talk of a parakeet in your home. Keep your bird in a decorative birdcage and let its song and chatter become a part of your home's familiar sounds.

Help the sick and tired.

*A*long with many illnesses come the sounds of coughs, sneezes, sniffles, blows, moans, and groans. Keep plenty of medicine on hand. To help clear your head when congested, steam the air with the soothing sound of a vaporizer. If you are sick and sleep with a mate, show

consideration by sleeping in another room. This will keep your spouse healthy and rested.

74

Have a "get-well bell."

Spoil a sick loved one with a designated "get-well bell" beside the bed. Whenever he or she needs something, a simple ring will let you know. This soothing sound and a little "TLC" is sure to help cure any mild illness.

75

Nod off naturally.

On a rainy night, open the windows and let the peaceful sound of the rain pouring soothe you to sleep. In the autumn, drift off to sleep listening to the rustling leaves falling from the trees. Let the sweet sound of birds singing outside your bedroom be your wake-up call.

76

Cut out creaking.

*I*f your mattress creaks or squeaks when
you roll over in it, it's time to replace it.
Purchase a quality, quiet mattress and box
spring set and get a good night's sleep.

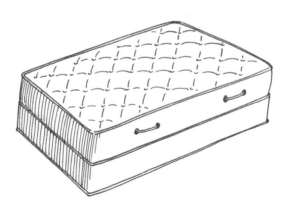

Shed tears and heartaches.

$\mathcal{S}$ome of our most painful, yet precious moments in life are when we have wept in the safety and comfort of our homes. Tears can be healing. Make your home a comforting place where family or friends feel the freedom to shed tears and heartaches.

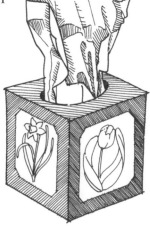

78

Drift into dreamland.

Have a clock radio with a sleep button. Set your radio on soft, soothing music and drift off to sleep. The sleep button will play twenty minutes or so of your favorite music before it shuts off automatically. If all goes well, you'll never hear it stop.

Tune it out.

*I*f there's no escaping distracting noises while you're trying to sleep, wear a pair of comfortable earplugs or mask the noise with a constant humming sound, such as an air conditioner, small fan, or space heater. Or try a sleep machine that continuously plays recordings of pleasant nature sounds to help you fall into a deep sleep.

80

Do not disturb.

*B*e sensitive and considerate to those sleeping. Tiptoe quietly. Whisper to others when nearby the sleeper. Turn off the telephone ringer in adjacent rooms. If you must wake up in the morning before your mate, lay out your clothes the night before and place them in the bathroom.

Do not Disturb

81

Forget the phone after bedtime.

*A*fter you have retired to your bedroom for the night, avoid taking phone calls. Talking on the telephone forces you to reenter the world, prevents you from mentally unwinding for a restful night's sleep, and keeps you from getting to bed on time. If someone calls in the middle of the night, consider it an emergency; if they call late in the evening, consider it impolite.

82

Wake up on the right side.

The way in which you wake up can set the tone for your whole day. Set your clock radio to wake you in the most pleasing way possible. Listen to a radio station that plays music you enjoy — if you're a light sleeper, keep the volume soft; if you're a heavy sleeper, turn it up loud. Set a CD-playing alarm clock so you can wake up with a grateful heart, listening to praise music.

Snore no more.

$\mathcal{I}$f you are a snorer, or sleep with a snorer, chances are you're not getting a restful night's sleep. For a quieter night, get a contoured snore pillow for the snorer. The foam pillow cradles the head and neck to allow proper breathing and peaceful sleeping. (If all else fails, investigate surgery to correct the snoring problem.)

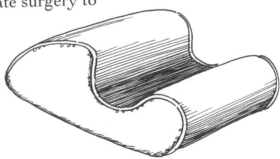

84

Be well-read.

Make it a family ritual to read aloud to one another. Read a Scripture or devotional thought at the breakfast table, a Bible verse to a loved one once you've crawled into bed, or an interesting article to stimulate discussion. As much as possible, read books to children or let them read them to you.

Good readers make leaders.

Look it up.

*D*on't miss a chance to expand your vocabulary. Keep a good dictionary and thesaurus beside your bed. When you are reading in bed and come across an unfamiliar word, look it up. Then begin using it in your vocabulary as often as possible.

86

Plan a peace talk.

Though it's best to settle disagreements before going to bed, there are those weary nights when further discussion only adds to the frustration. If this is the case, set a time and place for a peace talk. Once you are rested, refreshed, and have gained a proper perspective, resume your discussion and deal with the issue. Never avoid resolving an issue. A soothing home is a peaceful home.

Say a prayer.

$\mathscr{M}$ake it a nightly ritual to kneel at your young child's bedside for prayers, and also pray out loud beside or in your own bed before you go to sleep. Even a short prayer of thanks for the day lets God know you haven't forgotten his hand in your life.

88

Take time for tuck-in talks.

Cherish your children's tender years by treasuring the times that you tuck them into bed. Secure in the comfort and quiet of their beds, they are often willing to open their hearts to you. Listen, laugh, and shower them with love. Every now and then tape record one of your tuck-in talks and save it to listen to after they've become too big to tuck in.

89

Savor the sound of silence.

Every now and then, turn off every sound maker and enjoy the precious sound of silence. It will clear the air and your mind. You may even hear something you would have missed amidst the noise.

90

Treasure quiet time with God.

Guard your quiet time with God, as it sets the tone for your whole day. Get up early for time alone to read, write in a journal, or pray. The more time you spend with God in quiet solitude, the clearer you will hear his voice leading you through the busyness and noise of your day.

91

Stop the drip-drops.

Eliminate the irritating sound and waste of water by fixing leaky faucets and running toilets. If possible, learn a few plumbing tips and do simple repairs yourself.

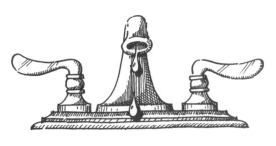

92

Hush the flush.

*F*or a quieter-sounding flush, buy or make a lovely fabric padded cover for your toilet seat lid. The fabric will absorb and soften the sound of flushing water. If you are purchasing or replacing a toilet, keep sound in mind as you make your selection.

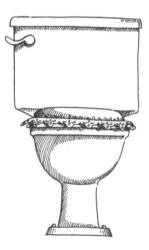

Fill your days with praise.

$\mathcal{M}$ake your home a place of praise. Listen to praise music as you get dressed in the morning. Play praise tapes or CDs and meditate on their worshipful words during your quiet time. Sing a praise song out loud when you feel it in your heart, or let one minister to your heart when your spirits need a lift.

94

Store it in your heart.

Store up the treasure of God's Word in your heart and mind by memorizing Scripture. Listen to Scripture memory melodies to learn verses. Hang a memory verse on your bathroom mirror and quote it out loud several times while getting ready for the day.

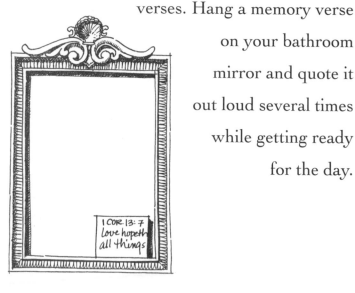

1 COR. 13: 7
Love hopeth
all things

Sing in the shower.

When standing under a sprinkling shower, go ahead and belt out a musical melody. Whatever others think doesn't matter. We all think we sound pretty good in the shower. Give your day a lift by lifting your voice.

96

Bring a song along.

*P*ortable tape or CD players travel with ease to your bedroom or bathroom. Select music appropriate to the mood you are trying to set.

Listen to romantic, peaceful, easy-listening music as you turn in for the day; upbeat music while getting dressed to go out on the town; fun music during children's bath time. Allow kids to keep a simple tape recorder in their room to listen to tapes while they play or rest.

Make music.

*I*f you or those you love enjoy playing an instrument, determine appropriate times and places for practicing. Whether it's the piano, guitar, flute, or other instrument, making music can make your home beautiful.

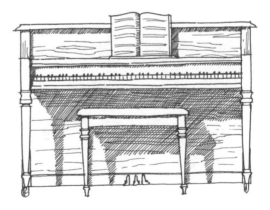

98

Wind up a memory.

*F*or a touch of beauty and music, place a pretty music box in a little spot where it will welcome others to wind it up. Hang a music mobile that plays a favorite lullaby from your baby's crib. The soothing sound will please baby's ears.

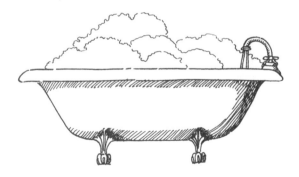

Wash away your cares.

Water relaxes us not only with its touch against our skin, but also with its sound to our ears. After a long, difficult day, turn on the bath water and listen to its peaceful sound as it rushes out the faucet. Let the water wash away your cares as it fills the tub.

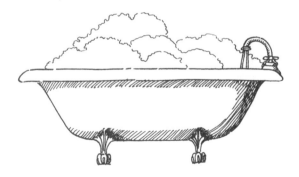

100

Tap, tap.

For a hospitable sound, mount a small door knocker on your powder room door. When guests fill your home, the light tap of the knocker will help determine if anyone is behind the closed door. Visitors will find the sweet touch delightful.

101

Tame your tongue.

Though every sound with which we fill our homes affects its atmosphere, none can be more soothing or saddening than our words. Spend time in God's Word studying the many Scriptures that refer to the powerful influence of the tongue. Use a concordance to help you locate appropriate passages. Ask God to help you build your home with loving and encouraging words.

More from Terry Willits . . .

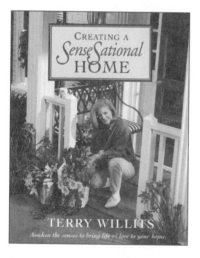

Creating a SenseSational Home is the complete guide to discover how awakening the five senses of sight, smell, taste, touch, and sound can create an atmosphere of love and cheer. From warmly-lit entrances that welcome family and friends to comfortable, homey interiors that invite them to stay and unwind . . . from fragrant bouquets to the tranquil ticking of a clock . . . *Creating a SenseSational Home* shows you simple and affordable ways to turn your home into a relaxing, inviting, and refreshing environment.

ISBN 0-310-20223-X
$19.99

ZondervanPublishingHouse
Grand Rapids, Michigan
http://www.zondervan.com

America Online
AOL Keyword:zon

A Division of HarperCollins*Publishers*